About the Author

Edilberto "Sandy" Santiago Jr. is a songwriter and blogger from Norzagaray, Bulacan, whose music celebrates honesty, kindness and community spirit. Raised in a humble agricultural family, he writes practical guides on farming techniques and survival skills tailored to the Philippine environment.

Passionate about arts and music, Edilberto uses his voice and his pen to inspire others—whether through uplifting verses or hands-on tips for life in the fields and the wild.

Permissions & Inquiries
Upper COC. Norzagaray
Bulacan 3002. Philippines
Email: edilberto@edilbertosantiago.com
Web: www.edilbertosantiago.com

TABLE OF CONTENTS

INTRODUCTION TO CASSAVA FARMING
Overview of Cassava

Cassava, also known as manioc or yuca, is a staple root crop that thrives in tropical and subtropical climates, making it well-suited for the diverse environments of the Philippines. This resilient plant is primarily cultivated for its starchy tuberous roots, which are a significant source of carbohydrates for millions of people worldwide. In the Philippines, cassava is not only a dietary staple but also an important cash crop that supports the livelihoods of many smallholder farmers. Its ability to grow in marginal soils and withstand drought conditions makes it an attractive option for home farmers looking to diversify their crops.

The cultivation of cassava in the Philippines typically begins with the selection of healthy planting materials, often referred to as stem cuttings. These stem cuttings should be obtained from high-yielding, disease-free varieties to ensure optimal growth and productivity. Home farmers should consider using organic methods for soil preparation, such as incorporating compost and organic fertilizers to enhance soil fertility. This organic approach not only promotes healthier plants but also contributes to sustainable farming practices by reducing reliance on chemical inputs.

As cassava plants grow, they require proper care to thrive. This includes regular watering, especially during dry spells, and the implementation of effective weed control measures. Organic mulching and intercropping with legumes can help suppress weeds while improving soil health. Pest management is another critical aspect of cassava farming, as pests like cassava mealybug and whitefly can severely affect crop yields. Sustainable pest management techniques, such as introducing beneficial insects and using organic pesticides, can help home farmers manage these threats while reducing their environmental impact.

Harvesting cassava is a labor-intensive process, typically occurring 8 to 12 months after planting, depending on the variety and growing conditions. Home farmers should monitor the tubers' size and maturity to determine the right time for harvest. Timing is crucial, as leaving cassava in the ground for too long can lead to reduced quality and increased susceptibility to pests. After harvesting, proper handling and storage of cassava roots are essential to prevent spoilage, especially in warm climates. Farmers can also explore value-added opportunities such as producing cassava flour or chips, which can enhance income potential.

In conclusion, cassava farming presents a viable opportunity for home farmers in the Philippines looking to engage in sustainable agricultural practices. By adopting organic farming techniques, implementing sustainable pest management strategies, and exploring value-added products, farmers can cultivate cassava effectively while contributing to food security and economic growth. As the demand for healthy, locally sourced food continues to rise, cassava could play a crucial role in transforming small-scale farming into a sustainable and profitable venture.

Importance of Cassava in the Philippines

Cassava, known locally as kamoteng kahoy, plays a crucial role in the agricultural landscape of the Philippines. As a staple food source, it provides an important alternative to rice and corn, particularly in rural areas where food security is a persistent concern. Due to its adaptability to various soil types and climate conditions, cassava can thrive in regions with limited agricultural resources. This resilience makes it an essential crop for home farmers looking to enhance their food production and resilience against economic fluctuations.

In addition to its nutritional value, cassava is a versatile crop that can be processed into various products, including flour, chips, and even biofuel. This versatility opens up numerous avenues for home farmers to diversify their income sources. By incorporating cassava into their farming practices, they can create value-added products that cater to local markets. This not only increases profitability but also encourages sustainable agricultural practices, as cassava can be grown alongside other crops, promoting biodiversity and improving soil health.

Organic cassava farming techniques have gained traction among home farmers who prioritize sustainability and health. By adopting organic practices, such as using natural fertilizers and pest repellents, farmers can cultivate cassava without the detrimental effects of synthetic chemicals. This approach not only benefits the environment but also appeals to health-conscious consumers seeking organic products. Implementing these techniques can lead to healthier crops, improved soil fertility, and reduced pest incidents, ultimately contributing to a more sustainable farming ecosystem.

Sustainable pest management is another vital aspect of cassava farming in the Philippines. Home farmers can employ integrated pest management strategies that focus on prevention and control rather than reliance on chemical pesticides. Techniques such as crop rotation, intercropping, and the use of biological control agents can significantly reduce pest populations while promoting a balanced ecosystem. By effectively managing pests, farmers can protect their cassava crops and ensure a bountiful harvest, which is essential for both food security and economic viability.

Lastly, cassava farming presents a promising small business opportunity for home farmers. With the increasing demand for cassava products, especially in urban areas, farmers can tap into lucrative markets by establishing local supply chains. By leveraging community support and local resources, they can create sustainable business models that not only provide financial benefits but also contribute to local economies. Emphasizing sustainable practices in their operations will further enhance their marketability, ensuring that cassava farming remains a viable and profitable endeavor for generations to come.

Benefits of Sustainable Farming Practices

Sustainable farming practices offer numerous benefits that can significantly enhance the productivity and resilience of cassava farming in the Philippines. One of the primary advantages is the improvement of soil health. Sustainable techniques such as crop rotation, cover cropping, and reduced tillage contribute to maintaining soil structure and fertility. By incorporating organic matter into the soil, farmers can boost microbial activity and nutrient availability, leading to healthier plants and increased yields. Healthy soil is essential for cassava, as it directly influences root development and overall crop vigor.

Another benefit of sustainable farming is the reduction of dependency on synthetic chemicals. Organic cassava farming techniques focus on natural pest management strategies, such as the use of beneficial insects, neem oil,

7

and organic repellents. These methods not only reduce the environmental impact associated with chemical fertilizers and pesticides but also promote biodiversity. By fostering a diverse ecosystem, farmers can create a more balanced environment that naturally controls pests and diseases, resulting in healthier crops and reduced production costs.

Sustainable farming practices also enhance water conservation, which is crucial in the Philippines' variable climate. Techniques such as mulching, contour farming, and the establishment of swales can help retain moisture in the soil and reduce erosion. These methods are particularly beneficial for cassava, which requires consistent soil moisture for optimal growth. By implementing water-saving strategies, home farmers can ensure their crops thrive even during dry spells, thereby securing a more stable income.

Moreover, sustainable cassava farming can open up new market opportunities. Consumers are increasingly seeking organic and sustainably produced foods. By adopting sustainable practices, farmers can differentiate their products in the marketplace, catering to the growing demand for organic cassava. This not only enhances the farmers' income potential but also contributes to the overall growth of the organic sector in the Philippines, positioning home farmers as key players in this burgeoning market.

Lastly, engaging in sustainable farming practices fosters a sense of community and stewardship among farmers. By sharing knowledge and techniques, home farmers can collaborate to create a network of support that enhances collective resilience against challenges such as climate change and market fluctuations. This cooperative approach not only empowers individual farmers but also strengthens the farming community, leading to greater sustainability and productivity in cassava farming as a small business opportunity. Embracing these practices is not merely about growing crops; it is about nurturing the land, community, and future generations.

GROWING CASSAVA IN THE PHILIPPINES
Selecting the Right Variety

Selecting the right variety of cassava is crucial for home farmers aiming to maximize yield and sustainability. The Philippines offers a diverse range of cassava varieties, each with its unique characteristics suited for different climates, soil types, and market demands. When choosing a variety, consider factors such as growth duration, disease resistance, and the intended use of the cassava, whether for food, animal feed, or as a raw material for processing. Engaging in local agricultural forums or consulting with agricultural extension services can provide valuable insights into popular and successful varieties in your specific region.

One of the primary considerations when selecting cassava varieties is the growth duration. Some varieties mature faster, typically within six to twelve months, making them ideal for farmers seeking quick returns on their investments. On the other hand, slower-maturing varieties may yield more substantial tubers, which can be beneficial for long-term sustainability. Home farmers should evaluate their specific needs, including market timing and personal consumption, to determine the best maturity period for their cassava crops.

Disease resistance is another critical factor in variety selection. Cassava is susceptible to various diseases, including cassava mosaic disease and bacterial blight, which can significantly affect yields. Choosing varieties that are resistant to these diseases can reduce the need for chemical interventions, aligning with organic farming practices. Farmers should seek out certified disease-free planting materials and may consider participating in community seed banks to access resistant varieties that have been well-adapted to local conditions.

In addition to disease resistance, understanding the end-use of cassava can guide variety selection. For home farmers interested in processing cassava into food products, certain varieties are known for their superior taste, texture, and nutritional value. Conversely, if the goal is to produce cassava for animal feed or industrial uses, different traits such as higher starch content may be prioritized. Assessing local market demands and preferences can help farmers choose varieties that not only thrive in their environment but also meet consumer expectations.

Finally, home farmers should also consider their own farming practices and resources when selecting cassava varieties. Sustainable pest management techniques, such as intercropping and organic pest control, can influence which varieties grow best in specific conditions. Farmers should evaluate their ability to manage pests and diseases organically and select varieties that align with their sustainable practices. By carefully considering these factors, home farmers can make informed decisions that enhance their cassava farming experience, promote sustainability, and potentially create a profitable small business opportunity.

Ideal Growing Conditions

The ideal growing conditions for cassava, particularly in the Philippines, are crucial for achieving optimal yields and ensuring the sustainability of your farming practices. Cassava thrives best in warm climates, making the tropical conditions of the Philippines particularly suitable. The ideal temperature range for cassava cultivation is between 25°C to 35°C. While cassava can tolerate some heat, it is important to monitor temperatures during extreme weather conditions, as prolonged exposure to temperatures above 40°C can stress the plants and reduce yields.

Soil quality plays a significant role in the successful growth of cassava. The plant prefers well-drained, sandy loam soils that are rich in organic matter. These soils allow for proper root development and facilitate the efficient uptake of nutrients. Soil pH is another important factor; cassava grows best in slightly acidic to neutral soils, with a pH ranging from 5.5 to 7.0. Home farmers should conduct soil tests before planting to determine the nutrient content and pH level, enabling them to amend the soil as necessary to create the best environment for their cassava crops.

Water management is essential for cassava farming. While cassava is drought-resistant, adequate moisture is vital during the early stages of growth to support root development. A consistent watering schedule, especially during dry spells, can make a significant difference in yield. Rainfall patterns in the Philippines can be unpredictable, so implementing sustainable irrigation techniques, such as drip irrigation or rainwater harvesting, can ensure that your cassava plants receive the necessary hydration without overwatering, which can lead to root rot and other diseases.

Pest and disease management is another critical aspect of creating ideal growing conditions for cassava. Home farmers should adopt integrated pest management (IPM) practices that combine biological, cultural, and mechanical controls to minimize the impact of pests while promoting a healthy ecosystem. Regular monitoring for pests, such as whiteflies and mealybugs, allows for early intervention. Planting resistant varieties and ensuring proper spacing between plants can also reduce the likelihood of pest infestations and disease outbreaks, leading to healthier crops and more sustainable farming practices.

Finally, proper crop rotation and intercropping can enhance the growing conditions for cassava. Rotating cassava with legumes or other crops helps to replenish soil nutrients and break pest and disease cycles. This practice not only contributes to the sustainability of your farming system but can also provide additional income streams by diversifying your produce. By understanding and implementing these ideal growing conditions, home farmers in the Philippines can cultivate resilient and productive cassava crops, paving the way for a successful small business opportunity in organic farming.

Planting Techniques

Planting techniques play a crucial role in the successful cultivation of cassava, particularly for home farmers in the Philippines. To ensure optimal growth, it is essential to select planting materials carefully. High-quality cassava cuttings should be chosen, ideally from healthy and disease-free plants. These cuttings should be approximately 20-30 centimeters long and taken from mature stems. Home farmers should aim to use varieties that are well-suited to local conditions, as these will often exhibit better resistance to pests and diseases, ultimately leading to higher yields.

The timing of planting is another significant factor in cassava cultivation. In the Philippines, the best planting season is usually at the onset of the rainy season, which typically begins around May or June. This timing allows the cassava to take advantage of the natural rainfall, reducing the need for supplemental irrigation. Home farmers should consider the local climate and soil conditions when planning their planting schedule, as these factors can greatly influence root development and overall crop performance.

When it comes to planting techniques, there are two primary methods that home farmers can employ: the traditional method and the ridge method. The traditional method involves planting cuttings directly into the soil, spaced adequately to allow for growth. On the other hand, the ridge method entails creating raised beds or ridges in the field, which can help improve drainage and soil aeration. This method is particularly beneficial in areas prone to waterlogging, as it minimizes root rot and other moisture-related issues. Home farmers should assess their specific site conditions and choose the method that best meets their needs.

Incorporating organic practices into planting techniques can enhance soil health and fertility. Before planting, home farmers can enrich the soil with organic matter, such as compost or well-rotted manure. This addition not only supplies essential nutrients but also improves soil structure and water retention, creating a more favorable environment for cassava roots to develop. Furthermore, using cover crops in the off-season can help suppress weeds and enhance soil quality, setting the stage for a successful cassava crop.

Sustainable pest management should also be integrated into the planting process. Home farmers can employ companion planting techniques, introducing pest-repelling plants alongside cassava to deter harmful insects. Regular monitoring of the crop for signs of pests or diseases is essential, allowing for early intervention when necessary.

By combining these planting techniques with sustainable practices, home farmers can create a resilient cassava farming system that not only yields high-quality produce but also contributes positively to the local ecosystem and serves as a viable small business opportunity.

Soil Preparation and Fertility

Soil preparation is a critical first step in cultivating healthy cassava crops, especially for home farmers in the Philippines. The process begins with a thorough assessment of the soil's physical and chemical properties. Testing the soil pH is essential, as cassava thrives in slightly acidic to neutral conditions, ideally between 5.5 and 7.0. Home farmers should consider using a soil testing kit to determine nutrient levels and pH. Based on the results, amendments such as lime or sulfur can be added to adjust the pH, while organic matter like compost can enhance soil structure and fertility.

Incorporating organic matter into the soil not only improves nutrient availability but also promotes beneficial microbial activity. This is particularly important in sustainable cassava farming, where the goal is to enhance soil health without relying on synthetic fertilizers. Farmers can utilize decomposed plant residues, animal manure, or green manure crops to enrich the soil. These organic amendments increase the soil's water-holding capacity and improve drainage, creating a more favorable environment for cassava roots to develop.

Fertility management is another crucial aspect of soil preparation. Cassava has specific nutrient requirements, particularly nitrogen, phosphorus, and potassium. To meet these needs organically, home farmers can implement a crop rotation system that includes leguminous plants, which naturally fix nitrogen in the soil. Additionally, applying organic fertilizers, such as fish emulsion or seaweed extract, can provide essential micronutrients. These practices not only sustain soil fertility but also reduce the reliance on chemical fertilizers, aligning with the principles of organic farming.

Pest management is integral to maintaining healthy crops and ensuring a successful harvest. Healthy soil contributes to robust plant growth, which in turn increases resistance to pests and diseases. Home farmers should adopt integrated pest management (IPM) strategies, such as introducing beneficial insects, using organic pesticides, and practicing crop diversity to minimize pest populations. Monitoring the health of the soil and plants can help identify potential issues early, allowing for timely interventions that do not compromise the organic integrity of the farming system.

Lastly, preparing the soil adequately sets the foundation for cassava farming as a viable small business opportunity. A well-prepared and fertile soil will lead to higher yields, which can significantly enhance profitability. Home farmers can capitalize on local markets by offering fresh, organic cassava products. By applying sustainable practices in both soil preparation and pest management, farmers not only contribute to their local ecosystems but also position themselves as responsible producers within the agricultural community. This holistic approach supports not just individual success but also broader efforts towards sustainable agriculture in the Philippines.

ORGANIC CASSAVA FARMING TECHNIQUES
Understanding Organic Farming Principles

Understanding organic farming principles is essential for home farmers looking to cultivate cassava sustainably in the Philippines. Organic farming emphasizes the use of natural processes and materials to enhance soil health, promote biodiversity, and reduce reliance on synthetic inputs. This approach aligns well with the unique conditions of cassava cultivation, as it enables farmers to create a resilient agricultural system that can thrive in the local environment while also contributing to the overall sustainability of their farming practices.

At the core of organic farming is the principle of soil health. Healthy soil is the foundation for robust cassava growth. Home farmers are encouraged to adopt practices such as crop rotation, cover cropping, and the use of organic amendments like compost and green manure. These strategies not only improve soil structure and fertility but also enhance microbial activity, which is crucial for nutrient cycling. By fostering a living soil ecosystem, farmers can ensure that their cassava plants receive the necessary nutrients for optimal growth without relying on chemical fertilizers.

Another critical principle of organic farming is biodiversity. By planting a variety of crops alongside cassava, home farmers can create a more resilient system that can better withstand pests and diseases. Intercropping with legumes or other complementary plants can improve soil nitrogen levels and discourage harmful insects. This diversity not only enhances pest management but also promotes a balanced ecosystem that supports beneficial organisms.

Implementing such practices can lead to healthier cassava plants and a more productive farming operation.

Sustainable pest management is a fundamental aspect of organic farming. Home farmers should focus on preventative measures, such as selecting disease-resistant cassava varieties and maintaining proper spacing to improve air circulation. Integrated Pest Management (IPM) strategies, which include biological control methods and organic pesticides, can also be employed. By monitoring pest populations and using natural predators, farmers can minimize crop damage while reducing the need for harmful chemicals. This holistic approach to pest management supports the health of both the crops and the surrounding environment.

Finally, organic farming principles can significantly enhance the economic viability of cassava farming as a small business opportunity. By producing organic cassava, farmers can tap into a growing market that values sustainably produced food. The increasing demand for organic products presents an opportunity for home farmers to differentiate their offerings and potentially command higher prices. By embracing organic practices, farmers not only contribute to their health and the environment but also create a sustainable livelihood that can benefit their communities. This alignment of ethical farming practices with economic incentives underscores the importance of understanding and implementing organic farming principles in cassava cultivation.

Organic Soil Management

Organic soil management is a fundamental aspect of sustainable cassava farming that ensures healthy crop growth and maximizes yield without relying on synthetic fertilizers. Home farmers in the Philippines can significantly benefit from understanding and implementing practices that enhance soil health. Healthy soil is rich in organic matter, which improves its structure, water retention, and nutrient availability. By focusing on organic amendments, farmers can create a thriving environment for cassava, allowing it to flourish even in less-than-ideal soil conditions. One effective method of organic soil management is the incorporation of organic matter, such as compost or well-rotted manure, into the soil. Composting kitchen scraps, farm residues, and green waste not only recycles nutrients but also enhances soil fertility. For cassava farming, it is essential to apply compost before planting, ensuring the soil is enriched with vital nutrients. Farmers can also use cover crops that fix nitrogen and add organic material to the soil, further promoting a healthy ecosystem. These practices not only improve soil quality but also reduce the need for chemical fertilizers, aligning with organic farming principles.

Crop rotation is another essential practice in organic soil management. By alternating cassava with other crops, farmers can disrupt pest and disease cycles while enhancing soil nutrient diversity. Leguminous plants, for instance, can be rotated with cassava to fix nitrogen in the soil, benefiting subsequent crops. This practice helps maintain soil fertility and reduces dependency on chemical inputs, making it a sustainable approach for home farmers looking to maximize their yields while minimizing environmental impact.

Sustainable pest management is closely linked to organic soil management. A diverse and healthy soil ecosystem promotes beneficial organisms such as earthworms and microbes, which can help control pests naturally. Farmers are encouraged to utilize integrated pest management (IPM) techniques, which involve monitoring pest populations and employing biological controls when necessary. By fostering a balanced soil environment, farmers can reduce the incidence of pests and diseases, ensuring healthier cassava plants and a more successful harvest.

Lastly, engaging in organic soil management not only contributes to healthier cassava crops but also enhances the overall sustainability of farming practices. Home farmers looking to turn cassava farming into a small business opportunity can leverage organic methods to appeal to a growing market of health-conscious consumers. By producing organically grown cassava, farmers can command higher prices and access niche markets, thus creating a sustainable income source. Investing in organic soil management is not just about improving crop yield; it is about building resilience in farming practices that benefit both the farmer and the environment.

Natural Fertilizers and Amendments

Natural fertilizers and amendments play a crucial role in sustainable cassava farming, particularly in the diverse agricultural landscape of the Philippines. Home farmers seeking to enhance their soil quality and promote healthy crop growth can benefit significantly from the use of organic materials. These fertilizers not only improve soil fertility but also help to maintain ecological balance and reduce dependence on synthetic chemicals. Incorporating natural amendments ensures that cassava plants receive essential nutrients while fostering a resilient farming ecosystem.

One of the most effective natural fertilizers for cassava is compost. Composting involves recycling organic waste materials such as kitchen scraps, plant residues, and animal manure. The decomposition process enriches the soil with vital nutrients, enhancing its structure and water retention capabilities. For home farmers, creating a compost pile is an accessible and eco-friendly method to produce high-quality fertilizer at little to no cost. By applying compost to the soil before planting cassava, farmers can provide a nutrient-rich foundation that supports robust plant growth and improves yield.

In addition to compost, other natural amendments like green manure and cover crops can significantly benefit cassava farming. Green manure, which consists of specific crops grown to be tilled back into the soil, adds organic matter and nutrients when decomposed. Common green manure crops include legumes, which fix nitrogen in the soil, thus improving its fertility. Likewise, cover crops such as peanuts or cowpeas can prevent soil erosion, suppress weeds, and enhance soil structure, creating an ideal environment for cassava cultivation. Integrating these practices into a farming routine can lead to healthier plants and increased productivity.

Moreover, using natural fertilizers can also play a vital role in sustainable pest management. Healthy soil leads to strong plants that are more resistant to pests and diseases. Home farmers can incorporate natural solutions like neem cake, which is derived from the seeds of the neem tree, to deter harmful insects while providing nutrients to the soil. Additionally, the use of mycorrhizal fungi can increase nutrient uptake for cassava plants, further enhancing their resilience against pests. By focusing on soil health and employing these natural amendments, farmers can reduce reliance on chemical pesticides and promote an environmentally friendly farming approach.

Ultimately, the integration of natural fertilizers and amendments in cassava farming not only contributes to improved soil health but also aligns with the principles of sustainable agriculture. Home farmers can create a productive and eco-conscious farming system that supports their livelihoods while protecting the environment. As cassava continues to emerge as a viable small business opportunity in the Philippines, adopting organic farming techniques will enable farmers to tap into the growing market demand for sustainably sourced agricultural products. Embracing these practices offers a pathway to thriving cassava production that benefits both the farmer and the ecosystem.

Crop Rotation and Intercropping

Crop rotation and intercropping are vital practices that can enhance the sustainability and productivity of cassava farming in the Philippines. These techniques involve strategically planning the planting of different crops in succession (crop rotation) or simultaneously (intercropping) to improve soil health, control pests, and optimize resource use. For home farmers looking to implement these practices, understanding the benefits and methods of crop rotation and intercropping can lead to healthier cassava plants and increased yields.

Crop rotation is the practice of alternating the types of crops grown in a specific area from season to season. This method helps to break the cycle of pests and diseases that can thrive when the same crop is planted repeatedly. For cassava farmers, rotating cassava with legumes or other crops like maize can enhance soil fertility, as legumes fix nitrogen in the soil. This natural fertilization reduces the need for chemical fertilizers, promoting organic farming techniques that are both environmentally friendly and economically beneficial for small-scale farmers.

Intercropping, on the other hand, involves growing two or more crops in close proximity during the same growing season. This practice can maximize land use and improve biodiversity on the farm. For cassava farmers, intercropping cassava with crops such as sweet potatoes or peanuts can provide mutual benefits. The taller cassava plants can offer shade to the shorter crops, while the legumes can improve soil nutrients, creating a symbiotic relationship that enhances overall crop health. Additionally, intercropping can lead to a more resilient farming system, as it diversifies the crops and reduces the risk of total crop failure due to pests or diseases.

Effective pest management is another significant advantage of both crop rotation and intercropping. By diversifying the crops in a field, farmers can disrupt the life cycles of pests that specifically target cassava. For instance, rotating cassava with crops that are less susceptible to certain pests can reduce their populations. Similarly, planting companion crops that attract beneficial insects can help in controlling pest populations without the need for chemical interventions. This aligns with sustainable pest management practices that are essential for organic cassava farming.

Incorporating crop rotation and intercropping into cassava farming not only promotes ecological balance but also opens up new small business opportunities. Farmers can diversify their produce, offering a range of crops in local markets and potentially increasing their income. By adopting these sustainable practices, home farmers can not only ensure a more productive and resilient cassava farm but also contribute positively to the environment and their local communities. Embracing crop rotation and intercropping can thus be a rewarding strategy for those looking to thrive in the cassava farming business in the Philippines.

SUSTAINABLE PEST MANAGEMENT FOR CASSAVA CROPS
Identifying Common Pests and Diseases

Identifying common pests and diseases is crucial for home farmers cultivating cassava in the Philippines. Cassava, while resilient, is vulnerable to a range of pests and diseases that can affect yield and quality. Familiarity with these adversities enables farmers to implement effective management strategies, ensuring healthier crops and a more sustainable farming approach. This subchapter will outline the most prevalent pests and diseases associated with cassava, providing essential identification tips and management practices.

One of the most common pests affecting cassava is the cassava mealybug. These small, white, cottony insects can weaken plants by sucking sap and excreting honeydew, which can lead to sooty mold growth. Signs of an infestation include stunted growth, yellowing leaves, and the presence of sticky residue on the plants. Additionally, cassava green mites may also pose a threat. These tiny pests cause leaf curling, discoloration, and defoliation, significantly impacting the plant's productivity. Regular monitoring of your crops can help catch these pests early, allowing for timely intervention.

Diseases can also pose significant challenges to cassava farmers. One notable disease is cassava mosaic disease, caused by viruses transmitted by aphids. Symptoms include yellowing and mottling of leaves, which can lead to reduced tuber yields. Another serious concern is cassava bacterial blight, which manifests as water-soaked lesions on leaves and stems, ultimately leading to plant wilting and death. Understanding the symptoms of these diseases is vital for farmers to take action before they spread throughout the crop.

Sustainable pest management techniques can be employed to combat these threats effectively. Integrated pest management (IPM) strategies, which include biological controls, cultural practices, and resistant varieties, can be particularly effective. For instance, introducing natural predators such as ladybugs can help control mealybug populations. Additionally, maintaining proper plant spacing and crop rotation can reduce pest infestations and disease outbreaks. Regularly inspecting plants and removing infected or infested specimens can further mitigate potential problems.

In conclusion, recognizing and addressing common pests and diseases is paramount for successful cassava farming in the Philippines. Home farmers must stay vigilant and informed about the signs of infestations and infections to employ timely management practices. By integrating sustainable pest management techniques, farmers can not only protect their crops but also create a more resilient farming system that contributes to the local economy and promotes organic practices. With the right knowledge and strategies, cassava farming can thrive as a sustainable small business opportunity.

Integrated Pest Management Strategies

Integrated Pest Management (IPM) strategies are essential for home farmers in the Philippines seeking to cultivate cassava sustainably. IPM combines various management approaches to control pest populations while minimizing environmental impact. It emphasizes understanding the pest life cycle, ecology, and their natural enemies, allowing farmers to make informed decisions that promote plant health and productivity. This holistic approach not only reduces reliance on chemical pesticides but also enhances the resilience of cassava crops against pest infestations.

One of the key components of IPM is regular monitoring and assessment of pest populations. Home farmers should inspect their cassava fields frequently to identify early signs of pest activity. Utilizing simple tools such as sticky traps or visual inspections can help in recognizing pest presence before they reach damaging levels. By keeping track of pest numbers and the specific species present, farmers can determine the most effective management tactics to employ, whether it involves cultural, biological, or mechanical methods.

Cultural practices play a significant role in managing pests in cassava farming. Proper crop rotation, intercropping with pest-repelling plants, and maintaining healthy soil through organic amendments can significantly reduce pest populations. For instance, planting legumes alongside cassava can enhance soil fertility and attract beneficial insects that prey on harmful pests. Additionally, managing planting dates and adjusting plant density can minimize pest pressure while optimizing growth conditions for cassava.

Biological control is another vital aspect of IPM that home farmers can implement. Introducing natural predators, such as ladybugs or parasitic wasps, can help regulate pest numbers without harming the ecosystem. Farmers can also utilize microbial pesticides, which are derived from natural sources and target specific pests. These biological agents are often safer for the environment and human health, aligning with the principles of organic farming. When combined with other IPM strategies, biological control can lead to sustainable pest management in cassava cultivation.

Finally, education and community engagement are crucial for successful IPM implementation. Home farmers should seek information from local agricultural extension services, workshops, and peer groups to stay updated on best practices and pest management innovations. Sharing experiences and solutions with fellow farmers can foster a collaborative approach to pest management. By embracing IPM strategies, home farmers in the Philippines can ensure the sustainability of their cassava crops, reduce costs associated with chemical inputs, and enhance the overall productivity of their small farming ventures.

Utilizing Beneficial Insects

Utilizing beneficial insects in cassava farming is an effective strategy for sustainable pest management. Home farmers can significantly reduce the reliance on chemical pesticides by encouraging the presence of natural predators and pollinators in their fields. Beneficial insects, such as ladybugs, lacewings, and parasitic wasps, play a crucial role in controlling pest populations that threaten cassava crops. These insects can help maintain a balanced ecosystem, ultimately leading to healthier plants and improved yields.

To attract beneficial insects, farmers can implement a variety of practices. Planting a diverse range of flowering plants around the cassava fields is one effective method. These plants provide nectar and pollen, which are essential food sources for adult beneficial insects. Additionally, creating habitats such as insect hotels or leaving patches of native vegetation can encourage the establishment of these natural allies. By fostering a welcoming environment for beneficial insects, home farmers can enhance their pest control strategies without resorting to chemical interventions.

Another critical aspect of utilizing beneficial insects is understanding the specific pests that threaten cassava crops. Common pests, such as aphids, whiteflies, and spider mites, can be effectively managed through the introduction of their natural enemies. For instance, ladybugs feed on aphids, while parasitic wasps can lay their eggs in harmful pest larvae, effectively reducing their populations. By identifying the right beneficial insects for the specific pests in their fields, home farmers can create a targeted and efficient pest management plan.

In addition to pest control, beneficial insects also contribute to the overall health of cassava crops through pollination. Some varieties of cassava benefit from insect pollinators, which can enhance seed production and increase yield. Encouraging a diverse insect population not only supports pest management but also promotes a more productive farming system. Home farmers should consider integrating practices that attract pollinators, such as planting flowering cover crops and minimizing disturbances to the habitat.

Finally, incorporating beneficial insects into sustainable cassava farming practices not only aids in pest management but also aligns with organic farming principles. This approach enhances the resilience of the farming system and can lead to improved soil health and increased biodiversity. As home farmers in the Philippines explore cassava farming as a small business opportunity, adopting these practices will not only yield higher profits but also contribute to the long-term sustainability of their agricultural endeavors. By embracing the role of beneficial insects, farmers can create a harmonious and productive environment for their cassava crops.

Organic Pest Control Methods

Organic pest control methods are essential for sustainable cassava farming, particularly in the diverse agricultural landscape of the Philippines. Home farmers are increasingly recognizing the importance of maintaining ecological balance while maximizing crop yields. By adopting organic pest control techniques, farmers can effectively manage pests without relying on synthetic chemicals, thus promoting a healthier environment and improving the quality of their cassava produce.

One of the most effective organic pest control methods is the use of natural predators. Introducing beneficial insects, such as ladybugs and lacewings, can help keep pest populations in check. These natural predators feed on common cassava pests like aphids and spider mites. Home farmers can enhance their gardens by creating habitats that attract these beneficial insects, such as planting flowering plants nearby to provide nectar and pollen. This approach not only helps control pests but also supports biodiversity in the farming ecosystem.

Another strategy is the application of organic pesticides made from natural ingredients. Solutions such as neem oil, garlic spray, and insecticidal soap can deter pests without harming beneficial insects. Neem oil, derived from the seeds of the neem tree, is particularly effective against a wide range of pests while being safe for plants and humans. Farmers should apply these organic solutions during the early morning or late afternoon to minimize the risk of harming beneficial insects and to enhance the effectiveness of the treatment.

Crop rotation is also a crucial component of organic pest management in cassava farming. By rotating cassava with other crops, farmers can disrupt the life cycles of pests that may have established in the soil. This method not only reduces pest populations but also improves soil health by preventing nutrient depletion. Home farmers should plan their planting schedules strategically, alternating cassava with legumes or other compatible crops to maintain soil fertility and reduce pest pressure.

Finally, maintaining healthy soil through organic practices can significantly enhance pest resistance in cassava plants. Healthy soil promotes robust root systems, which in turn lead to stronger plants that are less susceptible to pest infestations. Incorporating organic matter, such as compost or green manure, can enrich soil health and support beneficial microbial activity. Home farmers should prioritize soil health as a foundational aspect of their sustainable cassava farming practices, ensuring that their crops remain resilient against pests and diseases.

HARVESTING AND POST-HARVEST MANAGEMENT
Determining the Right Time to Harvest

The timing of cassava harvest is crucial for ensuring optimal yield and quality of the tubers. Home farmers in the Philippines should consider several factors when determining the right time to harvest. Generally, cassava is ready for harvest between 6 to 24 months after planting, depending on the variety and growing conditions. Farmers should closely monitor the growth stages of the plants and assess their readiness by observing the size and texture of the tubers. Harvesting too early can result in smaller, less starchy roots, while waiting too long can lead to fibrous, lower-quality tubers.

One effective method to ascertain the maturity of cassava is to examine the leaves. As cassava approaches maturity, the leaves will begin to yellow and drop, indicating that the plant is allocating energy to tuber development. Farmers should also inspect the tubers themselves by gently excavating a few to observe their size and firmness. A mature tuber should feel firm to the touch and be of a size that meets market standards. This hands-on approach allows farmers to make more informed decisions about the optimal harvest time.

Environmental factors also play a significant role in determining when to harvest cassava. The onset of dry weather can signal that it is time to harvest, as prolonged wet conditions can lead to rot and other diseases. Farmers should be aware of local weather patterns and plan their harvest accordingly. Additionally, soil conditions are important; overly wet soil can make harvesting difficult and may damage the tubers. Ensuring that the soil is adequately drained before harvesting can help maintain the integrity of the tubers.

For home farmers considering cassava farming as a small business opportunity, proper timing of the harvest can directly influence profitability. Harvesting at the right time can lead to better quality roots that attract higher market prices. Farmers should also consider the timing of their harvest in relation to local market demand. For instance, planning the harvest to coincide with local festivals or peak demand periods can maximize sales and profit margins. Understanding market trends and consumer preferences will enable farmers to position their products advantageously.

Lastly, sustainable practices should be integrated into the harvesting process. Using organic methods to manage pests and diseases during the growing period can lead to healthier plants and better yields. Farmers are encouraged to use hand tools rather than machinery to minimize soil compaction and preserve soil health. Furthermore, maintaining a record of harvest times and conditions can aid in future planning and improve overall efficiency in cassava farming. By implementing these strategies, home farmers can enhance their cassava production while contributing to sustainable farming practices in the Philippines.

Harvesting Techniques

Harvesting cassava requires careful planning and execution to ensure the highest yield and quality of tubers. The timing of the harvest is crucial; cassava can be harvested between 6 to 24 months after planting, depending on the variety and intended use. For home farmers, it is essential to monitor the growth of the plants closely. The right moment to harvest is when the leaves begin to yellow and fall off, indicating that the tubers have reached maturity. To maximize the quality of the cassava, farmers should aim to harvest when the tubers have reached a size suitable for their market or personal consumption.

When it comes to the actual harvesting process, it is important to use the right tools to avoid damaging the tubers. A spade or a digging fork is recommended for loosening the soil around the roots. Farmers should gently loosen the soil before pulling the tubers out to minimize the risk of bruising or breaking them. Careful handling is essential, as damaged tubers can lead to a decrease in quality and shelf life. Additionally, farmers should always wear gloves to protect their hands from the sap, which can cause irritation.

Post-harvest handling is just as important as the harvesting technique itself. Once the cassava is harvested, it should be promptly cleaned to remove any soil and debris. This step is crucial to prevent mold and decay during storage. After cleaning, the tubers should be sorted according to size and quality, with any damaged or diseased roots being set aside for immediate use or composting. Proper sorting not only helps maintain the quality of the product but also allows farmers to market the best tubers and maximize their profits.

Storage conditions play a key role in maintaining the quality of harvested cassava. The ideal storage environment is cool, dry, and well-ventilated. Home farmers can create suitable storage conditions by using simple techniques such as storing the tubers in wooden crates or baskets that allow air circulation. It is important to avoid direct sunlight, which can lead to spoilage. For those looking to extend the shelf life of their cassava, refrigeration can be an option, although it is not always practical for small-scale farmers.

Finally, farmers should consider value-added opportunities in cassava production. By exploring different processing methods, such as making cassava flour or chips, home farmers can diversify their product offerings and enhance their business potential. These value-added products often have a longer shelf life and can appeal to different market segments, including health-conscious consumers. Investing in education about processing techniques can empower farmers to transform their harvest into profitable small business opportunities while promoting sustainable practices in cassava farming.

Post-Harvest Handling and Storage

Post-harvest handling and storage are critical stages in cassava farming that significantly influence the quality and marketability of the crop. After harvesting, it is essential to minimize damage to the roots, as cassava is highly perishable. Farmers should aim to process cassava within 24 to 48 hours of harvesting to maintain its freshness. Upon harvesting, roots should be carefully dug out using a fork or spade, taking care not to bruise or cut them. This practice ensures that the roots remain intact, which is crucial for both storage and further processing.

Once harvested, the next step involves cleaning the roots to remove soil and debris. This should be done gently to avoid damaging the skin, which protects the root from microbial infections. Using a soft brush or cloth can help in this process. After cleaning, the cassava roots should be air-dried in a shaded area for a few hours. This not only reduces moisture content but also helps to prevent fungal growth during storage. Home farmers should be aware that roots stored with excess moisture are prone to rapid spoilage, potentially resulting in significant losses.

For effective storage, cassava roots should be placed in a cool, dry environment. Ideally, the storage area should be well-ventilated and away from direct sunlight. Using perforated bags or crates can facilitate air circulation, which is essential in prolonging the shelf life of the roots. Home farmers can also consider using natural preservatives, such as neem leaves, which have antifungal properties, to enhance the preservation of cassava during storage. Storing cassava roots at temperatures between 12 to 15 degrees Celsius can extend their viability for several weeks, making it easier to manage supply for both personal use and market sales.

Farmers interested in sustainable cassava farming should also consider the potential for value addition through processing. Techniques such as making cassava flour, chips, or even fermented products can not only improve the shelf life but also increase the economic value of the crop. Properly processed products can be stored for extended periods, providing an opportunity for home farmers to create a small business around their cassava production. This approach not only contributes to food security but also opens avenues for income generation, making cassava farming a viable small business opportunity.

Finally, it is crucial for home farmers to keep records of their post-harvest handling practices and storage methods. Documentation can help identify which techniques yield the best results and how different storage conditions affect the quality of cassava over time. By continually refining their methods based on past experiences, farmers can improve their overall productivity and profitability. Emphasizing sustainable practices in post-harvest handling and storage not only benefits the individual farmer but also contributes to the broader goal of promoting sustainable agriculture in the Philippines.

Processing Cassava for Market

Processing cassava for market involves several critical steps that ensure the product meets quality standards and is suitable for various consumer demands. Home farmers in the Philippines looking to tap into the market must understand the importance of proper harvesting techniques, as well as processing methods that enhance the shelf life and usability of cassava. Harvesting should be done when the roots are mature, typically 8 to 12 months after planting, to ensure optimal starch content. Farmers should carefully dig out the roots to avoid damage, as bruises can lead to spoilage and reduced market value.

Once harvested, cassava must be processed promptly to prevent deterioration. The first step in processing is thorough cleaning to remove soil and impurities. This can be done using water and a brush to ensure that the roots are free from contaminants. After cleaning, the cassava should be peeled to remove the toxic skin, which contains cyanogenic glycosides. This peeling process is crucial, as improper handling can lead to health risks for consumers. Home farmers may consider investing in proper peeling tools or techniques to enhance efficiency and ensure safety.

After cleaning and peeling, the next stage is slicing the cassava into uniform pieces, which facilitates even cooking and drying. Slicing can be done manually or with the help of slicing machines, depending on the scale of production. Once sliced, the cassava can be dried under the sun or using commercial dryers. Sun drying is a cost-effective method, but it requires good weather conditions, while dehydrators provide a more controlled environment, ensuring consistent moisture levels and reducing the risk of spoilage.

Post-drying, cassava can be processed into various value-added products, such as flour, chips, or even starch. These products not only cater to different market demands but also enhance profit margins for home farmers. Producing cassava flour, for instance, can open avenues in both local and export markets, especially with the increasing demand for gluten-free alternatives. Home farmers should consider investing in equipment that allows for efficient grinding and packaging to ensure their products are market-ready.

Finally, marketing plays a crucial role in the success of processed cassava products. Home farmers should engage in market research to identify potential buyers and distribution channels. Building relationships with local markets, restaurants, and online platforms can expand their reach. Additionally, emphasizing the organic and sustainable practices used in their farming can attract health-conscious consumers. By focusing on quality and effective marketing strategies, home farmers can create a sustainable business model centered around cassava processing, contributing not only to their income but also to the local economy.

CASSAVA FARMING AS A SMALL BUSINESS OPPORTUNITY
Understanding the Market for Cassava

Understanding the market for cassava is essential for home farmers in the Philippines who wish to maximize their yields and profitability. Cassava, a versatile root crop, has seen a growing demand in both local and international markets. Its various uses, ranging from food products to industrial applications, make it a valuable crop. Home farmers should familiarize themselves with the different segments of the market, including food processing, animal feed, and biofuel production. By understanding these market dynamics, farmers can make informed decisions on what varieties to grow and how to position their products.

In the Philippines, the demand for cassava as a staple food is increasing due to its affordability and nutritional value. With a growing population and a shift towards healthier eating habits, cassava is becoming a popular alternative to rice and other staples. Home farmers can tap into this market by producing high-quality cassava that meets consumer demands. It is also important to note that cassava flour is gaining traction in health-conscious markets, especially among those with gluten intolerance. By promoting organic farming techniques, home farmers can cater to this niche while enhancing the market appeal of their products.

Organic cassava farming techniques can significantly improve marketability and profitability. Consumers are increasingly seeking organic products, and cassava is no exception. Home farmers can adopt methods such as crop rotation, intercropping, and natural pest control to enhance soil health and crop resilience. These practices not only improve the quality of the cassava produced but also help in maintaining sustainable farming systems. Additionally, organic certification can provide farmers with a competitive edge in the market, allowing them to command higher prices for their produce.

Sustainable pest management is another critical aspect of understanding the market for cassava. Farmers must be aware of the common pests and diseases that affect cassava crops and how to manage them sustainably. Integrated pest management (IPM) techniques, including the use of natural predators, resistant varieties, and cultural practices, can minimize chemical inputs and promote healthier crops. By adopting these strategies, home farmers can ensure a consistent supply of high-quality cassava, which is vital for market success. Furthermore, educating consumers about the benefits of sustainable farming can enhance the reputation and demand for their products.

Finally, cassava farming presents a viable small business opportunity for home farmers in the Philippines. As the market for cassava continues to expand, farmers can explore various avenues for income generation, such as selling fresh roots, processed products, or even value-added items like cassava chips or flour. Establishing direct links with local markets, restaurants, and grocery stores can help farmers reach consumers more effectively. By understanding market trends and consumer preferences, home farmers can strategically position their cassava products, ensuring not only sustainability in their farming practices but also profitability in their business ventures.

Creating a Business Plan

Creating a business plan is a crucial step for home farmers looking to engage in sustainable cassava farming in the Philippines. A well-structured business plan serves as a roadmap, guiding farmers through the various stages of establishing and maintaining their cassava crops. It helps to clarify goals, outline strategies, and identify resources necessary for success. For those interested in growing cassava organically, the plan should detail methods for soil preparation, seed selection, and pest management techniques that align with sustainable practices.

The first component of a business plan involves conducting thorough market research. Understanding the demand for cassava, both locally and nationally, is essential. Farmers should identify potential buyers, such as local markets, restaurants, and food manufacturers, who are interested in purchasing organic cassava. Analyzing pricing trends and competition within the market can help farmers set realistic sales targets. This research will inform decisions on the scale of production and the potential profitability of the venture.

Next, the business plan should outline the production process. Home farmers need to detail the specific organic farming techniques they will employ, such as crop rotation, the use of organic fertilizers, and integrated pest management. Sustainable pest management is particularly important for maintaining healthy crops while minimizing environmental impact. Farmers can include strategies for monitoring pest populations and implementing biological controls, ensuring their practices are both effective and eco-friendly.

Financial planning is another critical aspect of the business plan. Farmers should project startup costs, including expenses for land preparation, seeds, organic fertilizers, and pest management supplies. Additionally, it is vital to estimate ongoing operational costs, such as labor and maintenance. Creating a budget and forecasting income based on market research will help farmers understand their break-even point and assess the viability of their cassava farming venture.

Finally, the business plan should include a section on growth and expansion. Home farmers should consider future opportunities, such as diversifying their product offerings by processing cassava into flour or snacks. They can also explore the possibility of community-supported agriculture (CSA) programs or partnerships with local businesses. By planning for growth, farmers can position themselves to adapt to changing market demands and increase their profitability in the long run.

Marketing Your Cassava Products

Marketing your cassava products effectively is essential for home farmers looking to turn their passion into a profitable venture. As the demand for organic and sustainably sourced food continues to rise, cassava presents a unique opportunity for farmers in the Philippines. Understanding the various marketing strategies available can help you reach your target audience, whether they are local consumers, restaurants, or health-conscious buyers.

One of the first steps in marketing your cassava products is to identify your target market. This involves researching the demographics and preferences of potential customers. For instance, urban areas might have a higher demand for organic produce, while local markets could be interested in traditional cassava products like flour or chips. By understanding who your customers are and what they seek, you can tailor your marketing efforts to meet their needs effectively.

Utilizing social media and online platforms is a powerful way to enhance your marketing efforts. Creating engaging content that highlights the benefits of your cassava products, such as their nutritional value and versatility in cooking, can attract attention. Platforms like Facebook and Instagram allow you to showcase your farming practices, share recipes, and connect with customers directly. Additionally, consider joining local online groups centered around organic and sustainable farming to network and share your offerings with a broader audience.

Participating in local farmers' markets and food fairs is another effective marketing strategy. These venues provide an excellent opportunity to showcase your cassava products, interact with potential customers, and receive immediate feedback. Engaging in community events not only helps build brand recognition but also fosters relationships with other local farmers and businesses. Collaborating with other vendors can lead to cross-promotional opportunities that expand your reach and customer base.

Lastly, consider establishing partnerships with local restaurants and health food stores that prioritize locally sourced and organic ingredients. Offering your cassava products to these establishments can create a steady demand and help you gain credibility in the market. Providing samples and educating potential partners about the benefits of cassava can encourage them to include your products in their menus or shelves. By diversifying your marketing approach and focusing on building relationships within your community, you can successfully promote your cassava products and grow your small business.

Financial Management for Small Farms

Financial management is a crucial aspect for small farms, particularly for those engaged in cassava cultivation in the Philippines. Understanding the financial landscape allows home farmers to make informed decisions that can lead to sustainable growth and profitability. It begins with establishing a clear understanding of the costs associated with cassava farming, which includes land preparation, seeds, fertilizers, pest management, and labor. By meticulously tracking these expenses, farmers can create a budget that reflects their operational needs and helps in setting realistic financial goals.

In the realm of organic cassava farming techniques, financial management also encompasses the evaluation of market prices for organic versus conventional cassava. Home farmers should conduct market research to understand the demand for organic cassava in their local areas and beyond. This knowledge not only assists in pricing their produce competitively but also helps in identifying potential buyers who are willing to pay a premium for organically grown cassava. Establishing connections with local markets or cooperatives can further enhance profitability and provide a stable income stream.

Sustainable pest management practices are another area where financial management plays a vital role. While investing in organic pest control methods may initially seem costly, the long-term benefits often outweigh these expenses. Home farmers should consider the cost-effectiveness of integrated pest management strategies that minimize chemical inputs while maintaining crop health. By calculating the potential return on investment for these sustainable practices, farmers can make more informed decisions that promote both environmental stewardship and financial viability.

Additionally, financial management should include an analysis of potential business opportunities related to cassava farming. This could involve diversifying income streams by exploring value-added products such as cassava chips, flour, or even biofuel. Home farmers can conduct feasibility studies to assess the market demand for these products and the costs involved in their production. By strategically planning for these opportunities, farmers can enhance their revenue potential and reduce reliance on a single crop.

Finally, maintaining accurate financial records is essential for the sustainability of small farms. Regularly updating income and expense reports allows farmers to assess their financial health and identify areas for improvement. Utilizing simple accounting tools or software can streamline this process, making it easier to track profitability and inform future decisions. By prioritizing financial management, home farmers can ensure that their cassava farming ventures are not only productive but also financially sustainable in the long run.

RESOURCES AND SUPPORT FOR CASSAVA FARMERS

Government Programs and Initiatives

Government programs and initiatives play a crucial role in promoting sustainable cassava farming in the Philippines. Recognizing the importance of cassava as a staple crop and its potential for economic growth, various government agencies have developed programs to support farmers in improving their practices. These initiatives aim to enhance productivity, ensure food security, and promote sustainable agricultural practices that protect the environment. Home farmers can benefit significantly from these programs, which often include training, financial assistance, and access to resources.

One of the key initiatives is the Department of Agriculture's (DA) Agricultural Training Institute (ATI), which offers training programs focusing on organic cassava farming techniques. These programs equip farmers with knowledge about sustainable farming practices, including soil management, crop rotation, and organic pest control methods. By participating in these training sessions, home farmers can learn how to cultivate cassava in an environmentally friendly manner, ultimately leading to healthier crops and a more sustainable farming system.

In addition to training, the government provides financial assistance through various grants and subsidies designed to support small-scale farmers. These funds can be used to purchase quality seeds, organic fertilizers, and necessary equipment for cassava farming. Home farmers can take advantage of these programs to reduce their initial investment costs, enabling them to focus on developing their business. By obtaining financial support, farmers can also explore value-added opportunities, such as processing cassava into flour or other products, which can further enhance their income.

Sustainable pest management is another critical focus of government initiatives related to cassava farming. The DA encourages the use of integrated pest management (IPM) strategies that combine biological, cultural, and mechanical methods to control pests and diseases affecting cassava crops. These strategies not only reduce the reliance on chemical pesticides but also promote a healthier ecosystem. Home farmers who adopt these sustainable pest management practices can protect their crops while minimizing environmental impact, making their farming operations more resilient and sustainable.

Lastly, government programs often emphasize the importance of farmer cooperatives and community-based organizations. By collaborating with local groups, home farmers can share resources, knowledge, and experiences, fostering a supportive network that enhances their farming practices. These cooperatives can also provide marketing support, helping farmers access larger markets and improve their income potential. By engaging with these government initiatives and community resources, home farmers can successfully navigate the challenges of cassava farming while contributing to a more sustainable agricultural landscape in the Philippines.

Non-Governmental Organizations and Support Groups

Non-Governmental Organizations (NGOs) and support groups play a crucial role in the development of sustainable cassava farming in the Philippines. These organizations are often at the forefront of agricultural innovation, providing farmers with access to resources, training, and technical assistance. They work to promote sustainable practices that not only enhance productivity but also protect the environment. By collaborating with local communities, NGOs can tailor their programs to the unique challenges faced by smallholder farmers, ensuring that their initiatives are both effective and culturally relevant.

One of the primary functions of NGOs in the cassava sector is to offer educational programs that focus on organic farming techniques. These programs teach farmers how to cultivate cassava without synthetic fertilizers and pesticides, thereby reducing chemical exposure and promoting soil health. Workshops often include hands-on demonstrations of organic pest management strategies, such as the use of beneficial insects and natural repellents, which are essential for sustainable pest control. This knowledge empowers farmers to make informed decisions about their farming practices, ultimately leading to healthier crops and improved yields.

Support groups, often formed by local farmers, provide an additional layer of assistance in the cassava farming community. These groups foster collaboration among farmers, enabling them to share experiences, resources, and best practices. By working together, farmers can tackle common challenges such as pest infestations, water scarcity, and market access. The collective knowledge within these groups can lead to innovative solutions that might not be possible for individual farmers working in isolation. Furthermore, these support networks can help farmers access financial resources or government programs designed to enhance agricultural productivity.

NGOs also play a significant role in promoting cassava farming as a viable small business opportunity. They often facilitate market linkages, helping farmers connect with buyers who are interested in sourcing locally grown cassava products. This is particularly important in the context of value-added products such as cassava flour, snacks, and biofuels, which can significantly increase farmers' income. By providing training in business management and marketing, NGOs empower farmers to transform their cassava production into sustainable enterprises that contribute to their livelihoods and local economies.

In conclusion, the collaboration between NGOs, support groups, and home farmers is essential for the advancement of sustainable cassava farming in the Philippines. These organizations not only provide valuable resources and training but also foster a sense of community among farmers. By embracing sustainable practices and leveraging collective knowledge, farmers can enhance their productivity, protect the environment, and explore new business opportunities, ensuring the long-term viability of cassava farming in the region.

Educational Resources and Workshops

Educational resources and workshops play a crucial role in empowering home farmers in the Philippines to successfully cultivate cassava. These resources provide essential information on the best practices for growing cassava, focusing on organic farming techniques that promote sustainability. Home farmers can access a variety of materials, including books, online courses, and local agricultural extension services, which offer tailored information specific to their regional conditions. By utilizing these resources, farmers can enhance their knowledge about soil health, crop rotation, and proper planting techniques, leading to more productive and resilient cassava crops.

Workshops are an invaluable tool for practical learning, allowing farmers to engage directly with experts and fellow growers. These events often cover a range of topics, such as organic pest management, soil fertility, and post-harvest handling. Participating in hands-on workshops can help farmers understand the nuances of sustainable practices, from identifying beneficial insects to using natural fertilizers. Additionally, workshops often provide opportunities for networking, where farmers can share experiences, challenges, and solutions, fostering a sense of community and collaboration among local cassava growers.

Sustainable pest management is a critical aspect of cassava farming that can be effectively addressed through educational initiatives. Home farmers can learn about integrated pest management (IPM) strategies, which combine biological, cultural, and mechanical methods to control pests while minimizing chemical inputs. Resources such as extension services and online platforms offer guidance on recognizing common cassava pests and their natural enemies. By adopting IPM practices, farmers can reduce their reliance on synthetic pesticides, thus promoting a healthier environment and improving the overall quality of their cassava produce.

Cassava farming also presents a viable small business opportunity for home farmers, which can be further enhanced through education and workshops. Farmers can learn about market trends, value addition, and processing techniques that can increase their profit margins. Educational resources often include information on business planning, marketing strategies, and access to financial support. By understanding the business aspects of cassava farming, home farmers can transform their agricultural practices into sustainable livelihoods, tapping into both local and export markets for cassava products.

In conclusion, the integration of educational resources and workshops into the cassava farming journey is essential for home farmers in the Philippines. These tools not only provide vital knowledge and skills but also foster a supportive community among farmers. By embracing sustainable practices, enhancing pest management techniques, and exploring business opportunities, home farmers can ensure the successful cultivation of cassava while contributing to food security and economic development in their regions. Through continued learning and collaboration, the future of cassava farming in the Philippines can thrive sustainably.

Networking with Other Farmers

Networking with other farmers is an essential component of successful cassava farming in the Philippines. Establishing connections with fellow farmers allows for the exchange of valuable knowledge, resources, and experiences. Home farmers can benefit significantly from participating in local agricultural groups, cooperatives, or online forums that focus on cassava cultivation. These platforms provide opportunities to share best practices, learn about organic farming techniques, and discuss sustainable pest management strategies that can enhance the productivity and health of cassava crops.

Attending local agricultural fairs and farmer's markets offers a chance to meet other cassava farmers and engage in discussions that can spark innovative ideas. These events often feature workshops and demonstrations on various topics related to cassava farming, including organic cultivation methods and pest control techniques. By participating in such gatherings, home farmers can gain insights into the latest trends in sustainable agriculture, while also establishing relationships that could lead to future collaborations or partnerships.

Joining farmer cooperatives or associations can be particularly beneficial for home farmers focusing on cassava as a small business opportunity. These organizations often provide members with access to shared resources, such as tools, seeds, and training programs. Cooperatives may also facilitate collective marketing efforts, allowing farmers to sell their cassava produce more effectively. By pooling resources and knowledge, members can enhance their business strategies, improve their market reach, and increase their overall profitability.

Networking with other farmers also fosters a community of support that can be invaluable during challenging times, such as pest outbreaks or adverse weather conditions. Sharing experiences and solutions can help home farmers develop effective management strategies that minimize losses and promote resilience. The exchange of information about sustainable pest management practices is particularly crucial, as it enables farmers to adopt environmentally friendly approaches that protect both their crops and the surrounding ecosystem.

In addition to local connections, leveraging online platforms can expand networking opportunities beyond geographical boundaries. Social media groups, agricultural blogs, and dedicated websites allow home farmers to connect with experts and peers from across the Philippines and beyond. Engaging in these online communities can lead to learning opportunities, mentorship, and even access to new markets for cassava products. By cultivating a diverse network of contacts, home farmers can enhance their knowledge and skills, ultimately contributing to the success of their cassava farming endeavors.

CASE STUDIES OF SUCCESSFUL CASSAVA FARMERS
Profiles of Innovative Farmers

In the Philippines, several innovative farmers have emerged as leaders in sustainable cassava farming, demonstrating how traditional practices can be adapted to modern needs. One such farmer is Maria Santos from the province of Quezon. Maria has successfully implemented organic farming techniques that not only enhance the health of her cassava crops but also improve the soil quality on her farm. By using compost made from kitchen scraps and local agricultural waste, she enriches her soil without resorting to chemical fertilizers. Maria's commitment to organic methods has significantly reduced her production costs while also increasing the marketability of her cassava as a premium product.

Another notable figure is Juan Reyes, who operates a small-scale cassava farm in Mindanao. Juan has embraced sustainable pest management strategies that minimize the use of harmful pesticides. He practices integrated pest management (IPM), which combines biological control methods with cultural practices to keep pest populations in check. For example, Juan introduces beneficial insects, like ladybugs, to combat aphid infestations. His approach has not only resulted in healthier crops but has also attracted local consumers interested in purchasing ecologically grown cassava, thereby enhancing his income.

In the Visayas region, we find Liza Delos Reyes, who has transformed her cassava farming venture into a thriving small business. Liza employs innovative processing techniques to add value to her cassava products, producing flour, chips, and other snacks that cater to health-conscious consumers. By diversifying her product line, she has tapped into a growing market for gluten-free and organic foods. Liza's story illustrates how home farmers can leverage cassava farming as a viable business opportunity, turning a simple crop into a source of sustainable income.

A key element of these farmers' success is their commitment to community engagement and knowledge sharing. Each of these farmers actively participates in local agricultural cooperatives, where they exchange best practices and resources with fellow farmers. For instance, Maria leads workshops on organic farming techniques, while Juan often shares his pest management experiences with new farmers in his community. This collective approach not only strengthens the farming community but also fosters a culture of innovation that benefits all members, encouraging sustainable practices across the board.

Finally, the profiles of these innovative farmers highlight the essential role of education and continuous learning in successful cassava farming. Each of these individuals has sought out training programs, workshops, and partnerships with agricultural organizations to enhance their knowledge and skills. This proactive approach to learning ensures they stay abreast of the latest sustainable farming techniques and market trends. As home farmers in the Philippines look to grow cassava, these profiles serve as inspiring examples of how innovation, community involvement, and education can drive success in sustainable agriculture.

Lessons Learned from Their Experiences

The journey of cassava farming in the Philippines has provided numerous lessons that can significantly benefit home farmers. One of the most critical lessons learned is the importance of soil health. Sustainable cassava farming begins with understanding the soil's composition and nutrient needs. Farmers have discovered that conducting soil tests can provide essential insights into pH levels and nutrient deficiencies. By amending the soil with organic matter, such as compost and green manures, farmers can improve soil fertility and structure, leading to healthier cassava plants and increased yields.

Another vital lesson is the value of crop rotation and intercropping. Many farmers have observed that rotating cassava with legumes or other crops helps break pest and disease cycles while enhancing soil nutrient levels. This practice not only promotes biodiversity but also ensures that the land remains productive over time. Intercropping cassava with crops like maize or vegetables can provide additional income streams while maximizing land use efficiency. Home farmers can adopt these practices to create a more resilient farming system that supports both their families and the environment.

Sustainable pest management has emerged as a cornerstone of successful cassava farming. Farmers have learned to identify common pests and their life cycles, enabling them to implement integrated pest management strategies effectively. This includes using natural predators, such as ladybugs and parasitic wasps, and employing organic pesticides made from neem oil or garlic. These methods have proven effective in minimizing pest damage while reducing reliance on chemical inputs. Home farmers can benefit from these strategies by creating a healthy ecosystem around their farms, thereby fostering a balance that reduces pest outbreaks.

Water management is another crucial lesson gleaned from the experiences of successful cassava farmers. The Philippines is prone to both drought and heavy rainfall, making effective water management essential for cassava production. Techniques such as mulching, using drip irrigation, and building swales can help conserve moisture and prevent soil erosion. Home farmers can adapt these practices to ensure their crops receive adequate water while minimizing waste, ultimately leading to healthier plants and higher yields. Understanding local weather patterns and planning accordingly can further enhance water management strategies.

Lastly, farmers have learned the importance of community and knowledge sharing in enhancing their cassava farming practices. Networking with local agricultural organizations, participating in workshops, and engaging with fellow farmers can provide valuable insights and support. Home farmers are encouraged to share their experiences and learn from one another, creating a collaborative environment that fosters innovation and sustainability. By building a strong community, farmers can collectively address challenges, celebrate successes, and ensure the long-term viability of cassava farming as a sustainable small business opportunity.

Adapting Strategies for Home Farmers

Adapting strategies for home farmers involves understanding the unique challenges and opportunities that come with cultivating cassava in the Philippines. Home farmers often operate on a smaller scale compared to commercial farms, which allows them to implement more personalized and sustainable practices. The first step in adapting strategies is selecting the right variety of cassava that is suited for local conditions. Farmers should consider factors such as soil type, climate, and resistance to pests and diseases. By choosing the appropriate variety, home farmers can enhance their yields and ensure the long-term health of their crops.

Organic farming techniques are essential for home farmers who want to grow cassava sustainably. Utilizing organic fertilizers, such as compost and green manure, can improve soil fertility without relying on chemical inputs. Additionally, practicing crop rotation and intercropping with legumes can help maintain soil health and reduce the risk of pest infestations. Home farmers should also consider using organic pesticides derived from natural sources, such as neem oil or garlic extracts, to manage pest populations effectively while minimizing environmental impact.

Sustainable pest management is a critical component of successful cassava farming. Home farmers can adopt integrated pest management (IPM) strategies that combine cultural, biological, and mechanical methods to control pests. Monitoring crops regularly for signs of pest activity is vital, as early detection can prevent severe infestations. Encouraging beneficial insects, such as ladybugs and lacewings, can provide natural pest control while promoting biodiversity on the farm. Establishing physical barriers, such as row covers, can also protect young cassava plants from pests without the need for chemical treatments.

Cassava farming presents a viable small business opportunity for home farmers in the Philippines. With the growing demand for cassava as a food source and industrial raw material, farmers can explore various value-added products, such as cassava flour, chips, and bioethanol. By diversifying their product offerings, home farmers can increase their income while reducing reliance on a single market. It is essential for farmers to conduct market research to identify local demand and potential buyers, which can help them tailor their production strategies accordingly.

In conclusion, adapting strategies for home farmers in cassava cultivation requires a commitment to sustainable practices and an understanding of local market dynamics. By focusing on organic farming techniques, sustainable pest management, and exploring business opportunities, home farmers can create a resilient and profitable farming operation. Continuous education and engagement with the farming community can also provide invaluable resources and support, enabling home farmers to thrive in the competitive agricultural landscape of the Philippines.

FUTURE OF CASSAVA FARMING IN THE PHILIPPINES
Trends in Sustainable Agriculture

Sustainable agriculture has gained significant traction in recent years, particularly as consumers and farmers alike become increasingly aware of the environmental impacts of conventional farming practices. In the context of cassava farming in the Philippines, sustainable agriculture emphasizes methods that maintain soil health, conserve water, and promote biodiversity. Home farmers are now adopting practices that not only enhance crop yields but also ensure that their farming methods are environmentally friendly. This shift is crucial for cassava, which is a staple food crop in the country and plays a vital role in the local economy.

One notable trend in sustainable cassava farming is the integration of organic farming techniques. Home farmers are turning to organic fertilizers and natural pest control methods to minimize the use of synthetic chemicals. Composting, for instance, has become popular as it enriches soil fertility while also recycling organic waste. Additionally, farmers are utilizing crop rotation and intercropping to break pest cycles and improve soil structure. These organic practices not only support the health of the cassava plants but also contribute to a more resilient farming system that can withstand the impacts of climate change.

Sustainable pest management is another essential trend that home farmers are embracing. Integrated Pest Management (IPM) strategies are increasingly being implemented to control pests in cassava crops without relying solely on chemical pesticides. This approach combines biological control methods, such as encouraging beneficial insects, with cultural practices like proper spacing and sanitation. Home farmers are also using pheromone traps and natural repellents derived from plants to manage pest populations effectively. By adopting these sustainable practices, farmers can protect their crops while minimizing harm to the environment and human health.

Water conservation methods are also becoming more prevalent among cassava farmers in the Philippines. Techniques such as rainwater harvesting, mulching, and drip irrigation are being implemented to optimize water use. These practices not only help in conserving precious water resources but also improve the resilience of cassava crops during dry spells. Home farmers are learning to implement these water-saving techniques to ensure their cassava plants thrive, even in unpredictable weather conditions. By focusing on efficient water management, farmers can enhance their productivity and sustainability.

Finally, the economic viability of cassava farming as a small business opportunity is gaining recognition. Sustainable agricultural practices not only lead to healthier crops but also open up new market opportunities for home farmers. Consumers are increasingly seeking organic and sustainably grown produce, creating a demand for cassava that meets these criteria. Farmers who adopt sustainable methods can tap into niche markets, potentially increasing their income while contributing to environmental conservation. As sustainable agriculture trends continue to evolve, home farmers in the Philippines can position themselves to benefit from the growing interest in responsibly produced food.

Potential for Export Markets

The potential for export markets for cassava in the Philippines is significant, given the increasing global demand for this versatile root crop. Cassava is not only a staple food in many regions, but it also serves as a key ingredient in various industrial applications, ranging from biofuels to food products. Home farmers who adopt sustainable practices in cassava cultivation can tap into these opportunities, providing them with both financial benefits and a chance to contribute to food security.

One of the most promising aspects of cassava as an export product is its adaptability to different climates and soils, which is particularly advantageous in the varied agricultural landscape of the Philippines. Home farmers can utilize organic farming techniques to enhance the quality of their cassava, making it more appealing to international buyers. Organic certification can further elevate the market value of their produce, as many consumers worldwide are increasingly seeking sustainably sourced products. By focusing on organic practices, farmers can not only improve their income but also promote healthier agricultural ecosystems.

In addition to food markets, there is a growing interest in cassava for industrial use, particularly in the production of starch, flour, and bioethanol. These products have a wide range of applications, including food processing, pharmaceuticals, and even biodegradable plastics. Home farmers can explore partnerships with local processing facilities or cooperatives to add value to their cassava crops. By becoming part of a larger supply chain, farmers can ensure that their products reach both domestic and international markets, thus expanding their business opportunities.

Sustainable pest management techniques play a crucial role in enhancing the export potential of cassava. By minimizing the use of chemical pesticides and adopting integrated pest management strategies, farmers can improve the quality and safety of their crops. This not only meets the increasing consumer demand for organic products but also reduces the risk of pest-related losses that can impact the quantity and quality of cassava available for export. Farmers who implement these practices are more likely to attract buyers who prioritize sustainability and safety in their sourcing decisions.

Finally, the government and various agricultural organizations in the Philippines are increasingly supportive of cassava farming initiatives aimed at export. Through training programs, financial assistance, and access to market information, these entities help home farmers navigate the complexities of entering export markets. By leveraging these resources, farmers can better position themselves to meet international standards and effectively market their cassava products. The combination of sustainable practices, organic farming, and support from agricultural institutions creates a robust foundation for home farmers looking to explore the potential of cassava in export markets.

Impact of Climate Change on Cassava Farming

Climate change poses significant challenges to cassava farming in the Philippines, affecting crop yields and the overall viability of this essential staple. Rising temperatures and altered rainfall patterns can lead to increased droughts or flooding, both of which adversely impact cassava production. Home farmers must understand these changes to adapt their farming practices effectively. The variability in climate can result in stunted growth, reduced tuber quality, and ultimately, lower harvests, making it imperative for farmers to stay informed about climate trends in their regions.

One of the most pressing effects of climate change on cassava farming is the increased prevalence of pests and diseases. Warmer temperatures can create favorable conditions for pests that threaten cassava crops, such as whiteflies and mealybugs. Additionally, shifts in climate may lead to the emergence of new diseases, further complicating efforts to maintain healthy crops. Sustainable pest management techniques become essential in this context, as they can help farmers mitigate the impact of these threats while minimizing chemical usage. Approaches such as biological control, crop rotation, and the use of resistant cassava varieties should be emphasized to ensure resilience against these challenges.

Water management is another critical area where climate change impacts cassava farming. The changing climate can lead to unpredictable rainfall patterns, making it challenging for farmers to ensure their crops receive adequate water. Drought conditions can severely limit cassava's growth, while excessive rainfall can lead to waterlogging and root rot. Home farmers can adopt various strategies to improve water management, such as implementing rainwater harvesting systems, using mulching techniques to retain soil moisture, or selecting drought-resistant varieties that can withstand periods of low water availability.

Soil health is also significantly affected by climate change. Increased rainfall can lead to soil erosion, nutrient leaching, and degradation of soil structure, all of which are detrimental to cassava farming. Farmers should focus on sustainable soil management practices, such as incorporating organic matter, using cover crops, and practicing conservation tillage to enhance soil fertility and structure. These practices not only help in maintaining healthy crops but also contribute to the overall sustainability of farming systems, making them essential for home farmers dedicated to organic cassava farming.

Finally, the economic implications of climate change on cassava farming cannot be overlooked. Fluctuations in yield due to climate variability can threaten the financial stability of home farmers who rely on cassava as a source of income. Understanding market dynamics and exploring value-added opportunities, such as producing cassava-based products, can provide farmers with additional resilience against economic shocks. By embracing sustainable practices and remaining adaptable to the ongoing changes in climate, home farmers can not only safeguard their livelihoods but also contribute to the sustainability of cassava farming in the Philippines.

Innovations in Cassava Farming Techniques

Innovations in cassava farming techniques have played a pivotal role in enhancing productivity and sustainability for home farmers in the Philippines. As the demand for cassava continues to grow due to its versatility and adaptability, adopting innovative practices can significantly improve yields while maintaining environmental stewardship. Home farmers can leverage new technologies and methods that align with organic principles, ensuring that their farming practices remain eco-friendly and economically viable.

One significant innovation in cassava farming is the use of improved seed varieties. Research institutions have developed high-yielding and disease-resistant cassava cultivars that are better suited to the Philippine climate. These varieties not only produce more tubers but are also more resilient to common diseases such as cassava mosaic disease and bacterial blight. Home farmers can obtain these improved seeds from local agricultural extension offices or cooperatives, ensuring they start their farming journey with the best possible planting material.

Another promising technique is the implementation of integrated pest management (IPM) strategies. Traditional pest control methods often rely heavily on chemical pesticides, which can be harmful to both the environment and human health. Innovations in IPM for cassava farming include biological control agents, such as introducing natural predators of pests, and employing physical barriers to prevent pest access. These sustainable pest management practices not only reduce chemical usage but also help maintain a balanced ecosystem, enhancing long-term soil health and crop resilience.

Soil health is crucial for successful cassava farming, and innovative practices such as agroecological approaches can greatly benefit home farmers. Techniques like intercropping with legumes can improve soil fertility and structure, while cover cropping can help prevent soil erosion and retain moisture. Additionally, the use of organic fertilizers, such as compost and green manures, can enhance soil microbial activity and nutrient availability. These practices not only support sustainable farming but can also lead to increased profitability, as healthier soils typically yield more robust crops.

Finally, the integration of technology in cassava farming presents unique opportunities for home farmers. Mobile applications and online platforms can provide farmers with real-time information on weather patterns, pest outbreaks, and market prices. Access to such information allows farmers to make informed decisions, optimizing their practices for better yields and profitability. Moreover, engaging with local farmer cooperatives can facilitate knowledge exchange, helping farmers adopt best practices and innovations collectively. By embracing these advancements, home farmers in the Philippines can transform cassava farming into a sustainable and profitable endeavor.

CONCLUSION
Recap of Key Points

The cultivation of cassava in the Philippines presents a significant opportunity for home farmers looking to engage in sustainable agriculture. This subchapter recaps the essential points discussed throughout the guide, emphasizing practical strategies for growing cassava in a way that is both environmentally friendly and economically viable. Understanding the specific climate and soil conditions in the Philippines is crucial, as cassava thrives in well-drained, sandy loam soils with adequate sunlight. Home farmers should ensure they select appropriate cassava varieties that are resilient to local pests and diseases while being suitable for the market demands.

Organic farming techniques play a pivotal role in sustainable cassava production. Home farmers are encouraged to implement practices such as crop rotation, intercropping with legumes, and the use of organic fertilizers like compost and green manure. These methods not only enhance soil fertility but also promote biodiversity, which is beneficial for pest management. The guide emphasizes the importance of nurturing the soil ecosystem to maintain a healthy crop yield without relying on synthetic chemicals, thereby aligning with organic farming principles.

Sustainable pest management is another critical aspect of successful cassava farming. The guide highlights integrated pest management (IPM) strategies that home farmers can adopt to minimize pest outbreaks while reducing the need for chemical pesticides. Techniques include monitoring pest populations, utilizing natural predators, and employing physical barriers to protect the crops. By fostering an environment where beneficial insects can thrive, farmers can create a balanced ecosystem that naturally regulates pest populations, ensuring the health of cassava plants over the long term.

Cassava farming offers a promising small business opportunity for home farmers in the Philippines. As demand for cassava products, such as flour and tapioca, continues to grow, farmers can capitalize on this trend by diversifying their offerings. The guide provides insights on value-added processing methods that allow farmers to transform raw cassava into market-ready products, increasing profitability. Additionally, establishing local markets and engaging in community-supported agriculture can help home farmers build a sustainable business model around their cassava crops.

In summary, this recap reinforces the key principles of sustainable cassava farming discussed throughout the guide. Home farmers are encouraged to focus on organic farming techniques, effective pest management, and market diversification to maximize their success. By adopting these practices, farmers not only contribute to environmental sustainability but also enhance their livelihoods through profitable cassava production. Engaging with local agricultural resources and communities can further empower home farmers to make informed decisions and foster a resilient farming system.

Encouragement for Home Farmers

Home farming offers a unique opportunity for individuals and families to engage directly with their food sources. For those considering or already involved in growing cassava, the potential for sustainable farming practices is immense. Cassava, a versatile and hardy crop, thrives in the varied climates of the Philippines, making it an ideal choice for home farmers. The cultivation of this root vegetable not only provides a reliable food source but also serves as a gateway to exploring organic farming techniques and sustainable practices that can enhance both yield and environmental health.

To successfully grow cassava, it is essential to understand the best planting practices and care techniques. Selecting quality cuttings, preparing the soil, and ensuring proper spacing are fundamental steps that can significantly influence the growth of your cassava plants. Home farmers should focus on organic methods, such as using compost and natural fertilizers, which can improve soil health and productivity. By adopting these practices, you not only cultivate more nutritious food but also contribute to a sustainable ecosystem that supports local biodiversity.

Sustainable pest management is another critical aspect of cassava farming that home farmers should prioritize. Rather than relying on chemical pesticides, consider integrating biological control methods, such as introducing beneficial insects that prey on common pests. Crop rotation and intercropping with other plants can also help in managing pest populations while enhancing soil fertility. By fostering a balanced ecosystem, home farmers can protect their cassava crops while minimizing environmental impact, creating a more resilient farming system.

As home farmers gain experience in cassava cultivation, they may discover opportunities to turn their passion into a small business. The demand for cassava in local markets continues to grow, driven by its versatility in various culinary applications and its use in processed products. By exploring value-added options, such as cassava flour or chips, home farmers can increase profitability and create a sustainable income stream. Educating oneself about market trends and consumer preferences can further enhance the chances of success.

In conclusion, the journey of a home farmer cultivating cassava in the Philippines is filled with opportunities for personal growth, environmental stewardship, and economic benefit. By embracing organic farming techniques, sustainable pest management practices, and exploring small business potentials, home farmers can make a significant impact on their communities and the environment. The path may present challenges, but with determination and a commitment to sustainable practices, cassava farming can yield not only a bountiful harvest but also a fulfilling lifestyle.

Vision for Sustainable Cassava Farming in the Future

The vision for sustainable cassava farming in the future emphasizes the integration of eco-friendly practices that not only enhance productivity but also ensure environmental health. Home farmers in the Philippines are uniquely positioned to adopt these practices, as they often have the flexibility to experiment with innovative techniques. By focusing on organic methods, these farmers can contribute to a more sustainable food system, reduce chemical inputs, and promote biodiversity. The future of cassava farming lies in a holistic approach that combines traditional knowledge with modern sustainable practices, enabling farmers to grow healthy crops while preserving their natural resources.

One key aspect of this vision involves the implementation of organic cassava farming techniques tailored to local conditions. Home farmers can utilize organic fertilizers, such as compost and green manure, to enrich the soil without relying on synthetic chemicals. Companion planting with legumes can enhance soil fertility and deter pests naturally. Additionally, the selection of disease-resistant cassava varieties can significantly reduce the reliance on pesticides, making the farming process safer for both farmers and consumers. By prioritizing organic methods, home farmers can produce high-quality cassava while fostering a healthier ecosystem.

Sustainable pest management is another critical component of the vision for the future of cassava farming. Home farmers must adopt integrated pest management (IPM) strategies that combine biological control, cultural practices, and mechanical methods to manage pests effectively. This approach minimizes chemical use while promoting natural pest predators and maintaining ecological balance. Regular monitoring of pest populations and crop health can help farmers make informed decisions and take timely action, ensuring that their crops remain healthy and productive. By embracing sustainable pest management, farmers can protect their livelihoods while safeguarding the environment.

Cassava farming also presents a viable small business opportunity for home farmers in the Philippines. With the growing demand for cassava products, including flour, chips, and bioethanol, farmers can tap into local and international markets. By adopting sustainable practices, they can differentiate their products, appealing to health-conscious consumers who seek organic options.

Establishing value-added processing units at home can further enhance profitability, allowing farmers to create a diverse range of cassava-based products. This entrepreneurial spirit not only supports individual farmers but also contributes to the economic development of their communities.

In conclusion, the vision for sustainable cassava farming in the future is one where home farmers play a pivotal role in transforming agricultural practices in the Philippines. By integrating organic farming techniques, sustainable pest management, and entrepreneurial opportunities, these farmers can cultivate cassava in a way that is both economically viable and environmentally responsible. As they embrace this vision, they will not only enhance their own livelihoods but also contribute to the broader goal of sustainable agriculture, ensuring that future generations can enjoy the benefits of healthy and resilient farming systems.

www.ingramcontent.com/pod-product-compliance
Lightning Source LLC
Chambersburg PA
CBHW070126230526
45472CB00004B/1445